the
sun's
birthday

the sun's birthday

john pearson

Doubleday & Company, Inc.   Garden City, New York

Also by John Pearson: TO BE NOBODY ELSE
KISS THE JOY AS IT FLIES

ISBN: 0-385-07412-3

Library of Congress Catalog Card Number 70-185426

Design by Jaren Dahlstrom, Crow-Quill Studios, San Francisco

First published in 1973 by Doubleday & Company, Inc.

GRATEFUL ACKNOWLEDGMENT IS GIVEN TO THE FOLLOWING FOR PERMISSION TO REPRINT:

From Catalog Nos. S-22 and S-26. Copyright by the California Institute of Technology and the Carnegie Institution of Washington. Photographs courtesy of The Hale Observatories.

"i who have died am alive again today and this". Copyright 1950 by E. E. Cummings. From "i thank you God", in his volume POEMS 1923-1954. Reprinted by permission of Harcourt Brace Jovanovich, Inc.

"no heart can leap, no soul can breathe". Copyright 1950 by E. E. Cummings. From "the great advantage of being alive" in his volume POEMS 1923-1954. Reprinted by permission of Harcourt Brace Jovanovich, Inc.

From THE FIRMAMENT OF TIME by Loren Eiseley. Copyright © 1960 by Loren Eiseley. Copyright © 1960 by The Trustees of the University of Pennsylvania. Reprinted by permission of Atheneum Publishers.

RANDOM HOUSE, INC.: for permission to quote from THE IMMENSE JOURNEY by Loren Eiseley. Copyright © 1946. Reprinted by permission of Random House, Inc.

This is a journey of the prowlings of one mind which has sought to explore, to understand, and to enjoy the miracles of this world.

from *The Immense Journey*
by Loren Eiseley

Sometimes the familiarity of life surrounding us creates a veil, and we cease to see it in its living, eloquent state. We need to borrow for a day the fresh vision of the photographer artist who by intent exploration, by keen attentiveness, by lingering meditation, rediscovers for us the life and illuminated beauty of each object of his magic lens.

A new universe emerges once more, in which the sparkle of a wave, the caress of light, the depth of a cave, the varied oriental design of fish and shell, the silk of dawns, move and breathe and touch us again, so that with them we ourselves come to life. The more he reveals, the more we see, the richer grows our universe, the deeper the challenge to love and participate and become at one with the feasts of color, of textures, of fluid moods, the fiestas and celebrations of our own life and nature's.

John Pearson has such a vision and the loving attentiveness which causes clouds to swirl, waves to emit light, and sand to carry messages. His vision offers us a way to commune with nature and the deepest rhythms in ourselves.

Anaïs Nin

i who have died am alive again today, and this
is the sun's birthday, this is the birthday of
life and of love and wings and the gay great
happening illimitably earth.

i am in love with this world, i have nestled
lovingly in it, i have climbed its mountains,
roamed its forests, sailed its waters . . .

. . . crossed its deserts, felt the sting of its frosts,
the oppression of its heats, the drench of its rains,
the fury of its winds, and always have beauty
and joy waited upon my goings and comings.

i dreamed that i floated at will in the great ether, and i saw the world floating also not far off, but diminished to the size of an apple. then an angel took it in his hand and brought it to me and said, "this must thou eat," and i ate the world.

no heart can leap, no soul can breathe
but by the sizeless truth of a dream . . .

. . . whose sleep is the sky and the earth and the sea.

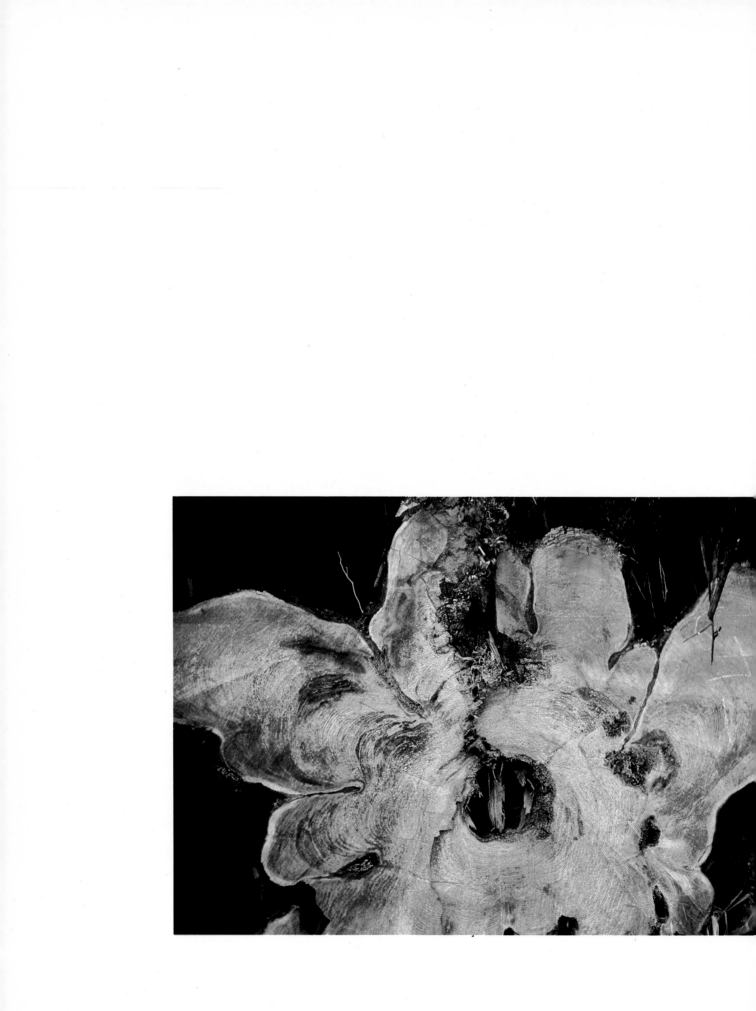

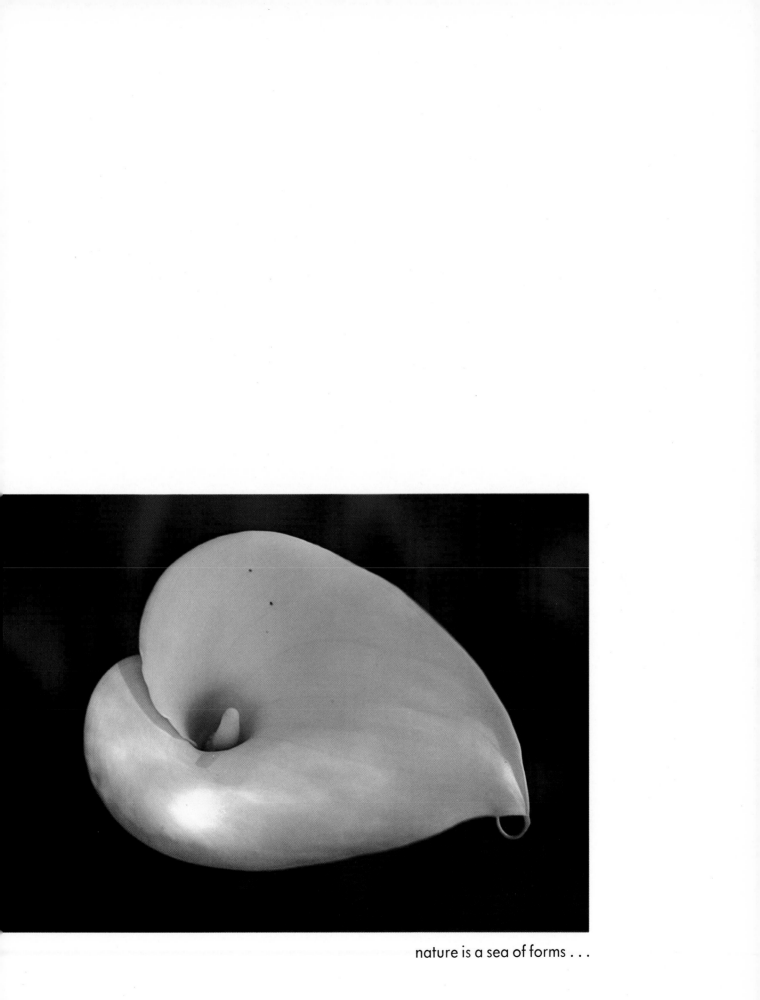

nature is a sea of forms . . .

. . . a leaf, a sunbeam, a landscape, the ocean — what
is common to them all is beauty . . .

. . . each moment has its own beauty; it
beholds every hour a picture which was
never seen before and which shall never
be seen again . . .

. . . every hour and season yields its tribute of delight — from breathless noon to grimmest midnight.

some of our richest days are those in which no sun shines outwardly, but so much more a sun shines inwardly.

behold these infinite relations, so like,
so unlike: many, yet one.

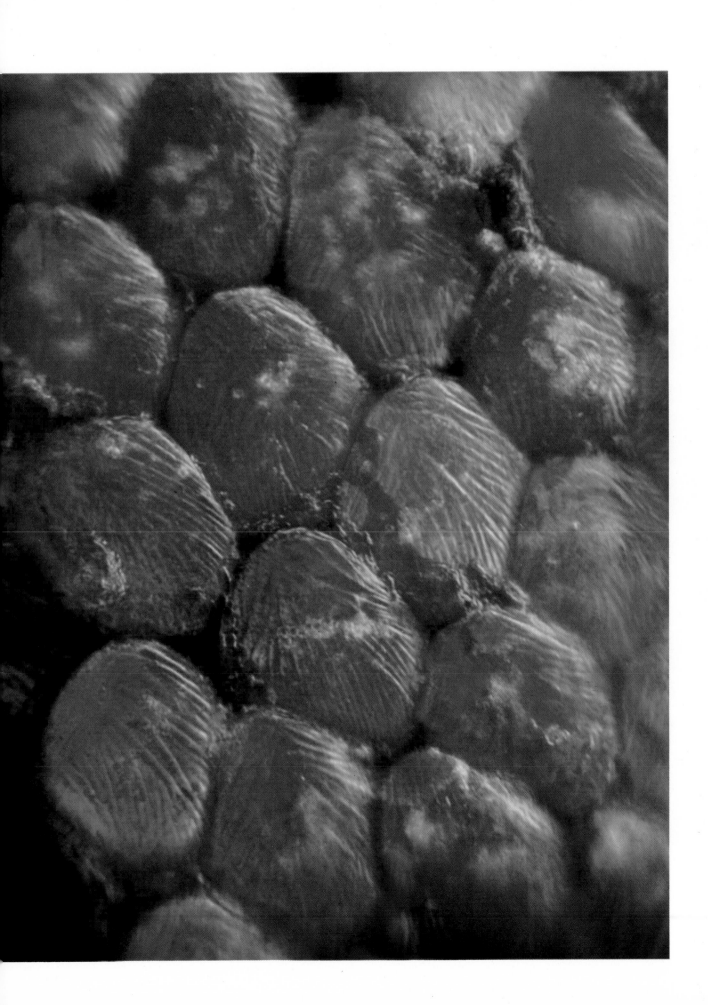

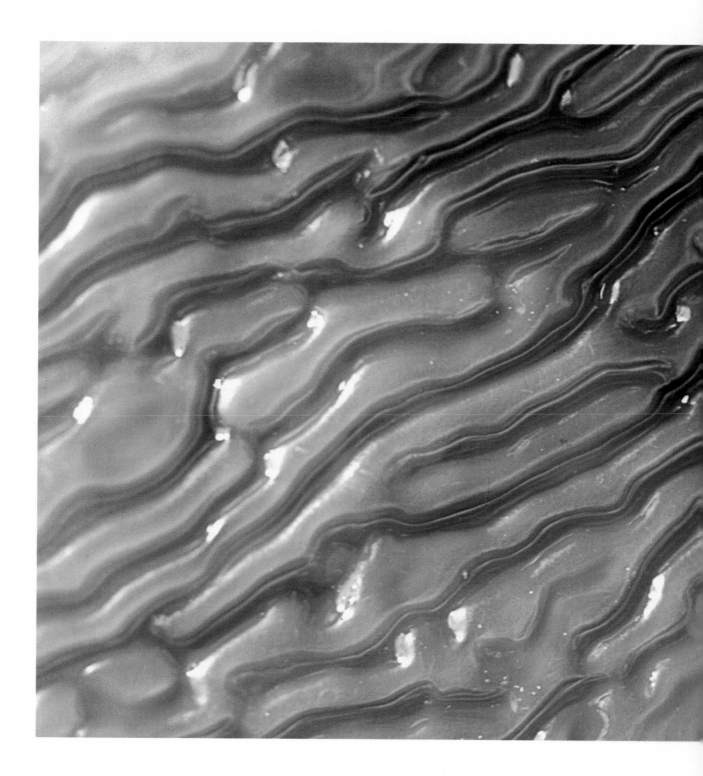

life is a train of moods like a string of
beads, and as we pass through them
they prove to be many colored lenses
which paint the world their own hue.

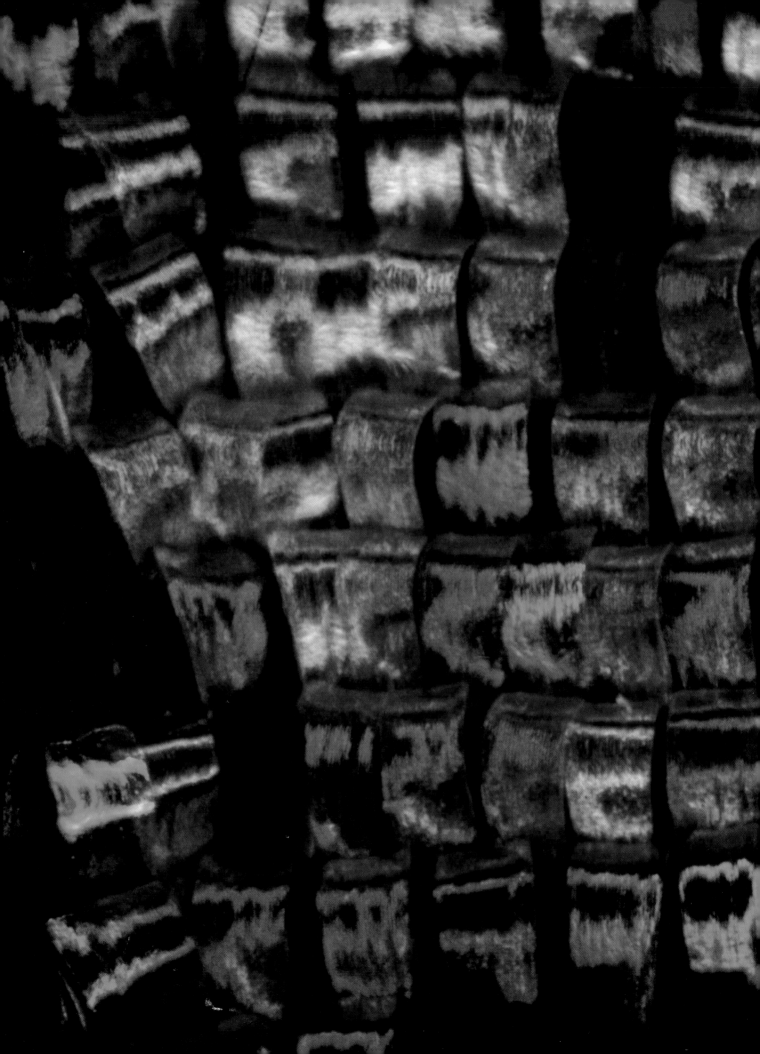

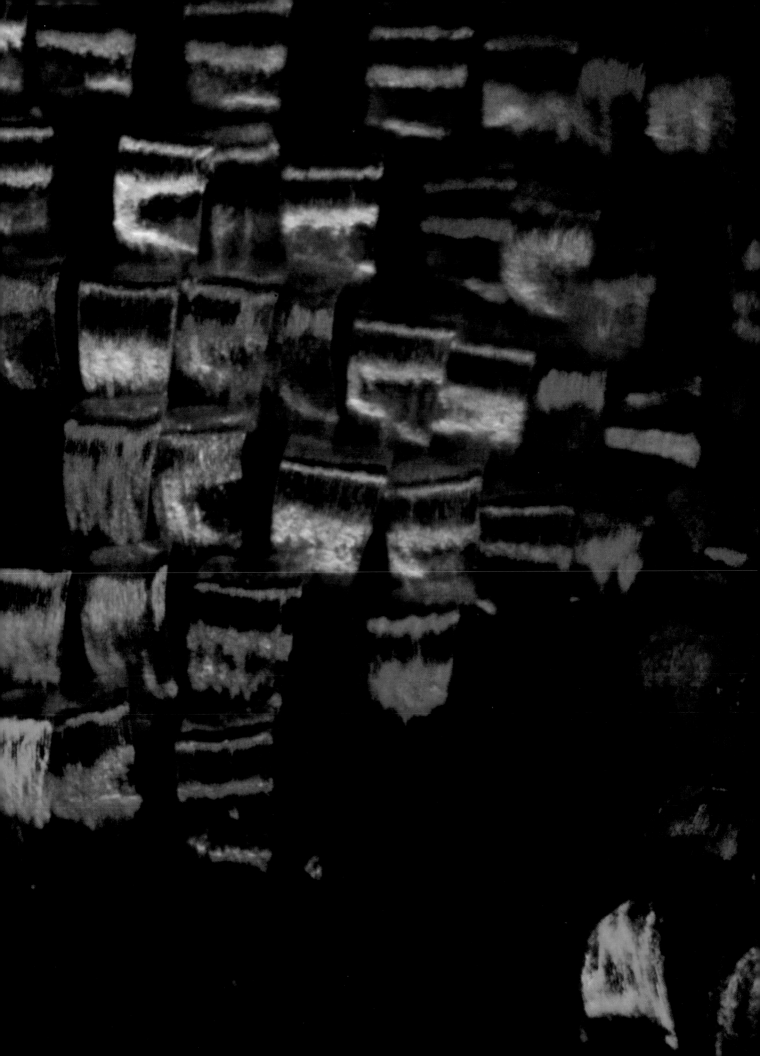

the whole of nature is a metaphor of
the human mind.

to see a world in a grain of sand and a
heaven in a wild flower, hold infinity in the
palm of your hand and eternity in an hour.

nature will bear the closest inspection: she
invites us to lay our eye level with the smallest
leaf, and take an insect view of its plain.
every part is full of life.

every leaf has its own tongue; every blade
of grass gives back its individual note . . .

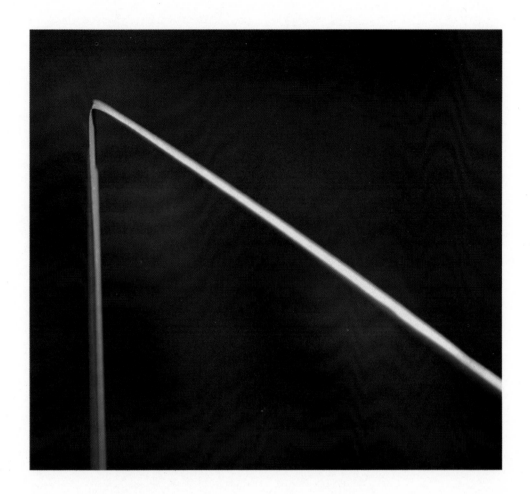

i believe a leaf of grass is no less than
the journey-work of the stars.

in the presence of nature a wild delight
runs through man, in spite of real sorrows.

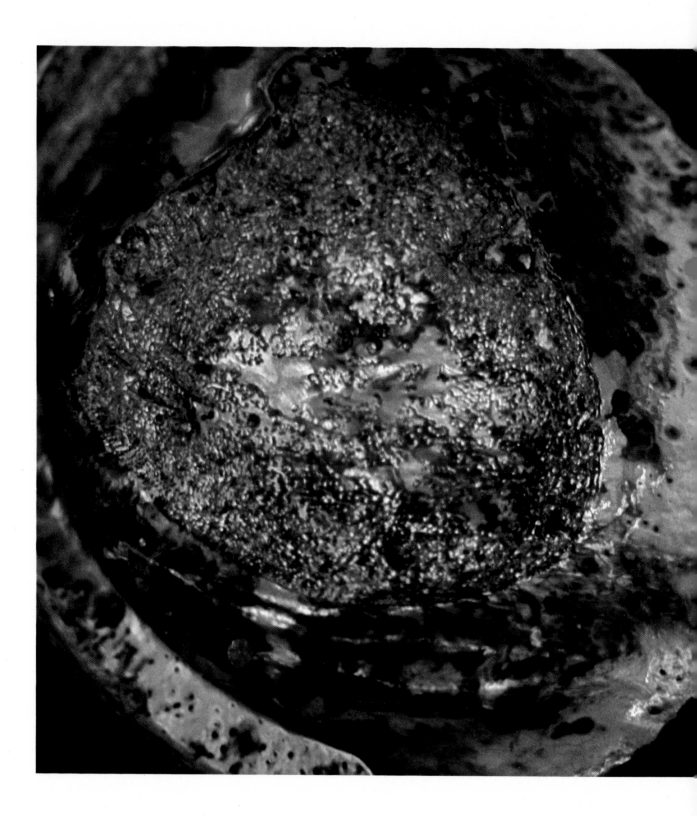

clouds come from time to time — and
bring to men a chance to rest from
looking at the moon.

once in a lifetime, if one is lucky, one so merges with sunlight and air and running water, that whole eons, the eons that mountains and deserts know, might pass in a single afternoon.

climb the mountains and get their
good tidings. nature's peace will flow
into you as sunshine flows into trees . . .

. . . the breath of the sea is built up into
new mountain ranges, warm during the
day, cool at night — good flower
opening weather.

once upon a time there were no flowers at all — only the green of a world whose plant life possessed no other color.

flowers changed the face of the planet. without
them the world we know — even man himself —
would never have existed . . .

. . . by a tenuous thread of living protoplasm, we are linked forever to lost beaches . . .

. . . the stars have shifted far or vanished
in their courses, but still the naked glistening
thread winds onward . . .

... no one knows the secret of its beginning or its end ...

. . . its forms are phantoms . . .

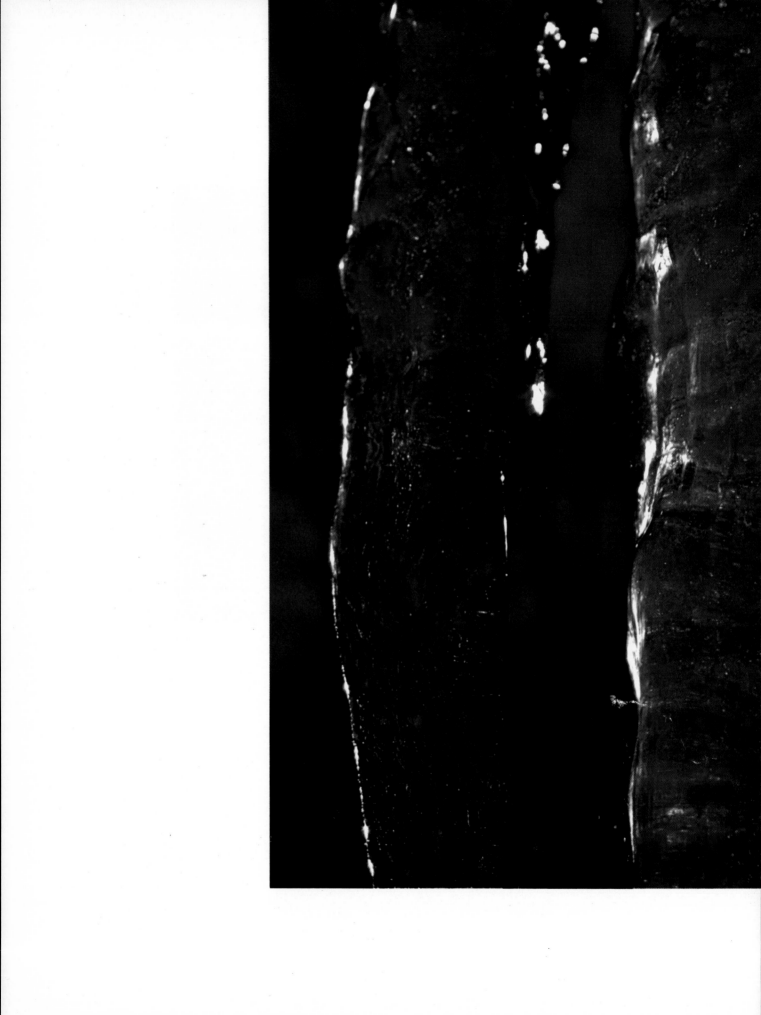

. . . it has appearances, but at its heart lies water which brings into being nine tenths of everything alive.

turtle and fish and the pinpoint chirpings
of frogs are all watery projections, as
man himself is a concentration of salt and
sun and time.

will you seek afar off? you will come back at
last to things best known to you, finding
happiness, knowledge, not in another place,
but in this place — not for another hour,
but this hour.

out of the sane, silent, beauteous miracles
that envelop and fuse me — trees, water,
grass, sunlight, and early frost — the one
i am looking at most today is the sky . . .

. . . the vacant spaciousness of air . . .

. . . and the veiled blue of the heavens seem miracles enough.

an early twilight ushers in a long
evening in which many thoughts have
time to take root and unfold.

a lake is earth's eye, looking into which
the beholder measures the depth of
his own nature.

nature will have clefts where i may hide,
and sweet valleys in whose silence
i may weep undisturbed . . .

she will cleanse me in great waters and make me whole.

the true harvest of my daily life is
intangible as the tints of morning or
evening. it is a little stardust caught,
a segment of the rainbow.

as long as i live, i'll hear waterfalls and
birds sing, and get as near the heart of
the world as i can.

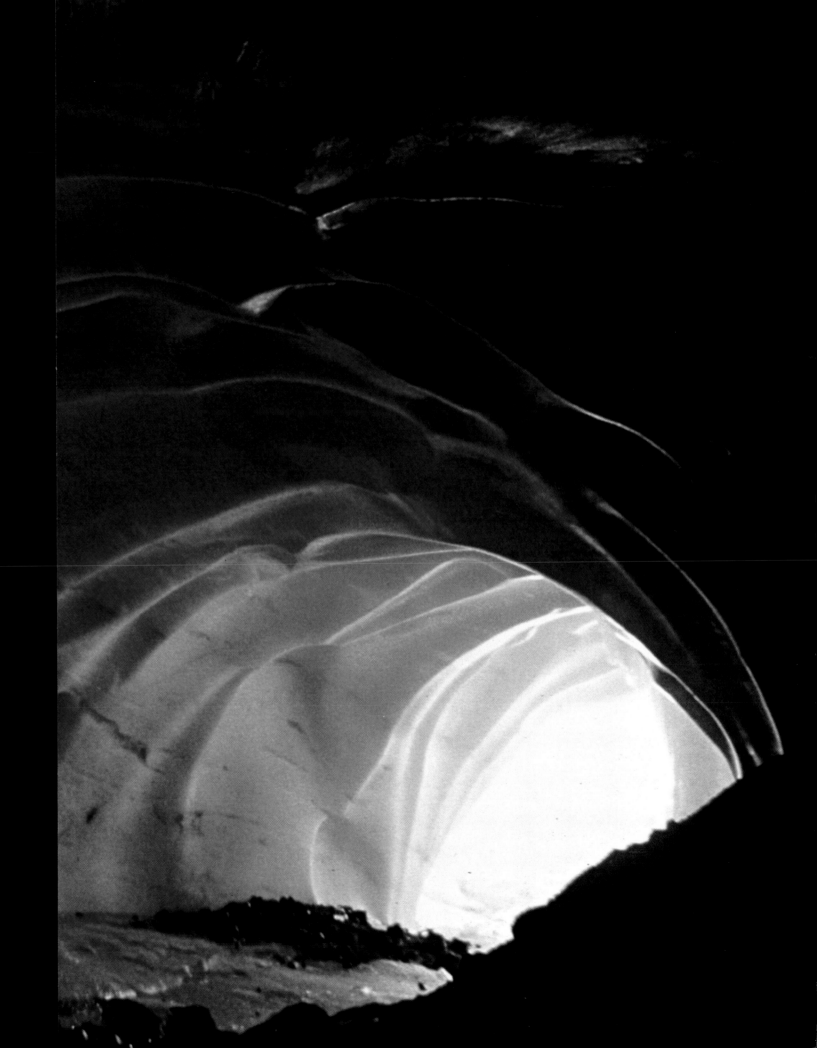

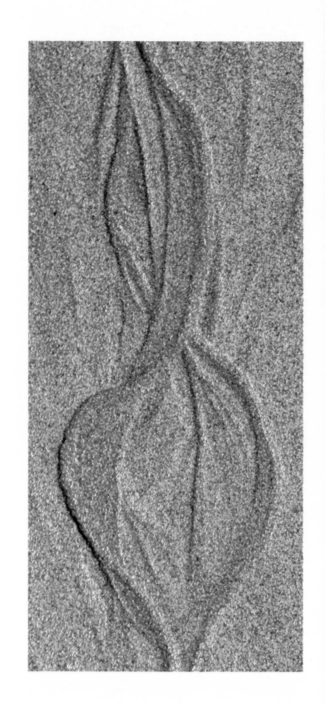

dwell as near as possible to the channel in which your life flows.

if the doors of perception were cleansed,
everything would appear to man as
it is — infinite . . .

. . . infinite . . .

The quotations are from the following sources:

Page

8	e. e. cummings, *Collected Poems*
12	John Burroughs
14	John Burroughs
16	Ralph Waldo Emerson
18, 19	e. e. cummings, *Collected Poems*
21, 24, 27, 30	Ralph Waldo Emerson
32	Henry David Thoreau
34	Ralph Waldo Emerson
38	Ralph Waldo Emerson
42	Ralph Waldo Emerson
44	William Blake
47	Henry David Thoreau
48	François Chateaubriand
51	Walt Whitman
52	Ralph Waldo Emerson
56	Matsuo Basho
58	Loren Eiseley, *The Immense Journey*
61, 63	John Muir
64, 67	Loren Eiseley, *The Immense Journey*
68, 69, 70, 71	Loren Eiseley, *The Firmament of Time*
73, 76	Loren Eiseley, *The Immense Journey*
78	Walt Whitman, *Leaves of Grass*
80, 82, 83	Walt Whitman, *Specimen Days*
84, 86	Henry David Thoreau, *Walden*
91	Oscar Wilde
94	Henry David Thoreau, *Walden*
96	John Muir
98	Henry David Thoreau, *Journal*
100	William Blake

Sources of photographs other than the author's:

16   Photo used with permission of California Institute of Technology and Carnegie Institution of Washington.

35, 36, 40-41   Micro-photographs by Stennett Heaton, Heaton's Biological Laboratory, San Leandro, California. I want to express my appreciation to Stennett for the use of three of his incredibly beautiful micro-photographs of insects. His work and spirit are pure poetry.

69   Photo used with permission of California Institute of Technology and Carnegie Institution of Washington.

JOHN PEARSON—Born 1934. . .minister's son. . .the South. . .protection, love, trust. . .
one older sister. . .routine hangups. . .college, graduate school. . .much information,
little understanding. . .marry at twenty-five. . .move to California. . .Karen is born
. . .minister two years. . .think about nuclear war. . .poverty. . .What can I do?. . .
Who am I?. . .quit. . .move to Berkeley. . .unemployed. . .two degrees—no job. . .
office worker. . .intense searching, boredom, despair. . .separation, divorce,
depression, tranquilizers, analysis, shock therapy. . .thirty. . .forget the princess,
embrace the dragons, feel the sun, taste the ocean, let go. . .miracle. . .re-birth. . .hack
photographer. . .weddings, little leagues, sororities. . .a bummer. . .but it's O.K.. . .all
inner now. . .first good job. . .freedom, trust, expansion, MONEY. . .photograph what
I love and feel in free time. . .help from friends, photographers, unexpected places. . .
thirty-one. . .live alone and start first book. . .thirty-two. . .MEET LIZ. . .someone to
share it all with. . .her place. . .my place. . .three weeks. . .our place. . .closer. . .
deeper. . .begin to know my daughter. . .first read Anais Nin. . .incredible. . .beautiful
. . .true. . .How does she know?. . .thirty-four. . .*To Be Nobody Else* published. . .many
help, care, make it possible. . .camping in Europe. . .new people. . .new nature. . .
thirty-five. . .return to Berkeley. . .see what is close, near overlooked. . .read Anais Nin's
third Diary. . .read first two again. . .photograph. . .read. . .talk. . .listen. . .print
pictures from Europe. . .expansion. . .exhaustion. . .sleep. . .more awareness of
dreams. . .changes. . .work mostly at night. . .Liz brings coffee. . .everyone close helps
. . .reactions, suggestions, revelations. . .fourteen hours in the darkroom. . .more
changes. . .*Kiss the Joy as It Flies* comes together. . .What next?. . .Nothing. . .relax
. . .try to do nothing. . .nothing is impossible!. . .another book?. . .re-read Whitman,
Emerson, Thoreau, John Muir. . .wish they were alive. . .feel lost. . .nature is
inexhaustible. . .many trips to the Kodak lab. . .*The Sun's Birthday* takes shape. . .a
few months later, almost published. . .glad it wasn't. . .changes, refinements. . .new
words, new photographs, new experiences. . .travel in the Everglades and the
Okefenokee swamp. . .return to Berkeley. . .more trips to Kodak. . .more late nights. . .
love Jaren's design, it's becoming a book. . .Liz brings coffee, scrambled eggs and
bacon. . .a year later. . .we mail the final revision to Doubleday. . .celebration. . .
rejoicing. . .Liz brings Irish coffee. . .now thirty-seven. . .The future. . .14,000 days left
. . .more music. . .more poetry. . .more dreams. . .more dragons. . .more art
. . .more nature. . .more life. . .

How to thank everyone? So many helped. LIZ LAMSON—made everything possible,
believed in me. MY PARENTS—who have always believed. JAREN DAHLSTROM—did
the beautiful layout and design, endlessly generous with his time and talents. DON
GERRARD—friend, editor, publisher—a hundred things. TOM BAIRD—taught me the
craft of photography and helped me trust my own vision. JOE EHRLICH—published the
first book and encouraged the other two. CHUCK MORRELL—loved *The Sun's Birthday*
even in its first tenuous form—two 5 x 7 plastic wedding albums. His care and genuine
enthusiasm led to its becoming a reality.

Photo by Karen Pearson